Selection and Use of Replacement Methods in Animal Experimentation

FRAME

Fund for the Replacement of Animals in Medical Experiments

Russell & Burch House, 96-98 North Sherwood Street, Nottingham NG1 4EE

UFAW

Universities Federation for Animal Welfare

The Old School, Brewhouse Hill, Wheathampstead, Herts AL4 8AN

ISBN 0 900767 98 7

Contents

1. PREFACE

Those applying for a project licence under the *Animals (Scientific Procedures) Act 1986* are required to sign a statement declaring that they have considered the feasibility of achieving the purpose of the project by means not involving regulated procedures on animals under the Act, and to state that in their opinion no such alternatives would achieve the objectives of their project. European legislation also requires the proper consideration of alternatives to animal experimentation, and the principle of REDUCTION as one of Russell and Burch's 3Rs is now well established in the scientific community.

This booklet is a joint production by FRAME and UFAW. The aim is to provide a practical guide to help ensure that those considering animal experimentation have explored all opportunities to avoid animal use and attempted to minimise the numbers involved. The booklet is only the first port of call, but it does highlight the different categories of replacement techniques, and provides references and addresses for further information. New possibilities for replacement will continue to be developed and it is the responsibility of those carrying out animal experiments to remain up-to-date in this area.

2. INTRODUCTION AND AIMS

Replacement is just one of the 3Rs ('reduction, refinement and replacement'.) proposed by Russell and Burch in *The Principles of Humane Experimental Technique* (1959), now reprinted and available from UFAW. The 3Rs approach to experimentation aims to achieve best scientific practice whilst implementing, as far as possible, a reduction in the numbers of animals used and refinements in experimental techniques, such that any pain, suffering or distress caused to the animals is minimised. Where possible, replacement techniques should always be used, provided that the scientific objectives can still be attained.

When completing a project licence application form, investigators are required by law to declare that they have given full consideration to the possible use of replacement methods. However, there is no readily available guidance to assist the applicant in meeting this legal requirement. These guidelines aim to improve the conduct of research and testing, and the achievement of research objectives by focusing on best scientific practice and on potential alternative methods. In addition, they will help investigators to meet their legal obligation to consider fully and competently alternatives to the use of animals.

As animal procedures are used in such a wide range of scientific areas, it is impossible to describe the possible replacement methods available to all scientists in all fields of work. Therefore, this booklet aims to give a summary of the current uses, advantages and limitations of some non-*in vivo* methods, so that the scientist can use this information as a first stage in a search for possible alternatives. Key references are provided, so that the user can investigate the possibility for applying each type of alternative method. In addition, the booklet provides a guide to other sources of information on alternatives.

A scheme in the Appendix provides scientists with a structured strategy which should be followed when deciding whether any alternatives can be implemented in their studies. The strategy takes into account the potential for the total or partial replacement of animals, and the possibility of achieving reduction and refinement. This scheme should be followed by all scientists when planning a research or testing strategy and protocol. The scheme could also be used by the Home Office and local ethics committees when assessing the necessity for using animals in a particular project.

3. LEGISLATIVE REQUIREMENTS

The UK *Animals (Scientific Procedures) Act 1986* (Anon 1986a) contains several important clauses relating to animal experimentation. Clause 5(4) states:

> ***'In determining whether and on what terms to grant a project licence, the Secretary of State shall weigh the likely adverse effects on the animals concerned against the benefit likely to accrue as a result of the programme of work to be specified in the licence.'***

Clause 5(5) states:

> ***'The Secretary of State shall not grant a project licence unless he is satisfied that the applicant has given adequate consideration to the feasibility of achieving the purpose of the programme to be specified in the licence by means not involving the use of protected animals.'***

Such protected animals are defined in the Act as all living vertebrate animals, except man, as well as one invertebrate species, the common octopus *(Octopus vulgaris)*, and includes foetal, larval or embryonic forms, which have attained specific stages in their development.

Lastly, Clause 5(6) states that:

> ***'The Secretary of State shall not grant a project licence authorising the use of cats, dogs, primates or equinae, unless he is satisfied that animals of no other species are suitable for the purposes of the programme to be specified in the licence, or that it is not practicable to obtain animals of any other species that are suitable for those purposes.'***

Hence, there is consideration given in the Act to increased protection for so-called companion species.

In Europe, *Directive 86/609/EEC* (Anon 1986b) and the *Council of Europe Convention for the Protection of Vertebrate Animals Used for Experimental and Other Scientific Purposes* (Anon 1986c) both require that:

> ***'An experiment shall not be performed if another scientifically satisfactory method of obtaining the result sought, not entailing the use of an animal, is reasonably and practicably available.'***

The concept of replacements to the use of protected animals is, therefore, promoted via UK and European legislation.

In the UK, applicants for project licences under the 1986 Act are required to sign a declaration (in Section 21 of the application form) stating that they have considered the possibility of replacing animal use with methods 'not involving regulated procedures on animals protected under the Act', and that no such replacements 'would achieve the objectives of the project.'

4. INTRODUCTION TO THE THREE Rs

Many scientists assume that the concept of 'alternatives' relates to the use of techniques which completely replace the use of animals in research. They will subsequently dismiss the idea that alternatives could be implemented into their work, as their work involves the use of *in vivo* methods. However, the term 'alternatives' refers not only to replacement methods, but encompasses all the 3Rs (reduction, refinement and replacement) of Russell and Burch (1959). They defined **reduction** as a means of lowering 'the number of animals used to obtain information of a given amount and precision', **refinement** as any development leading to a 'decrease in the incidence or severity of inhumane procedures applied to those animals which have to be used', and **replacement** as 'any scientific method employing non-sentient material which may, in the history of animal experimentation, replace methods which use conscious living vertebrates'. The importance of their approach lies in its combination of animal welfare considerations with good science and best practice.

It is important to understand that replacement, reduction and refinement alternatives should not be considered in isolation from one another, and that each can affect the others. For example, the use of non-animal methods does not only offer the possibility of reducing the numbers of animals that might have to be used subsequently, for example, when screening candidate chemicals during the early stages of product development, but can also lead to refinement of such animal experiments, when they are necessarily used. In addition, each of the 3Rs should only be applied when attainment of the scientific objective of the proposed, and worthwhile, investigation will not be compromised.

It is customary for new alternative assays to undergo some form of validation process. Validation refers to the process whereby the reliability and relevance of an assay are established for a particular purpose. The scale and complexity of validation is therefore dictated by the purpose of the assay. For example, where a method is to be used for in-house purposes only, then validation can be conducted on a limited scale. If an assay is designed to replace an animal method and be used widely for regulatory purposes (for example, safety testing), then validation trials are often conducted at several different laboratories according to specific internationally harmonised criteria, which are intended to facilitate regulatory acceptance of the alternative method. Validation has been the subject of considerable discussion (Balls *et al* 1995a; Anon 1996; Balls & Fentem 1997).

The objectives of a recent ECVAM (European Centre for the Validation of Alternative Methods) workshop were to discuss the current status of the 3Rs, and to make recommendations aimed at achieving greater acceptance of the concept of humane experimental technique and the more-active implementation of reduction alternatives, refinement alternatives and replacement alternatives (Balls *et al* 1995c). The report of this workshop is essential reading for all scientists, before commencing any project in which the use of animals is planned.

4.1 Refinement

Refinement encompasses those methods which alleviate or minimise potential pain and distress, and which enhance animal well-being. The implementation of refinement techniques can be simple, (for example, improvements in housing conditions), and the importance of refinement should not be overlooked when planning an experimental protocol. It is not possible to cover the concept of refinement in this booklet and the reader is referred to several recommendations, reviews, and articles on aspects of refinement (Morton *et al* 1993a,b; Flecknell 1994; Morton 1995; Poole 1997). In addition, the journal *Laboratory Animals* frequently publishes articles on the refinement of experimental procedures.

4.2 Reduction

The term 'reduction' describes strategies for obtaining comparable levels of information from the use of fewer animals in scientific procedures, or for obtaining more information from a given number of animals, so that, in the long run, fewer animals are needed to complete a given research project or test. There are areas of overlap between reduction and replacement. For example, *in vitro* methods are frequently used in the initial screening of compounds for toxicity, so that the most toxic compounds can be rejected before continuing with further tests in animals. The *in vitro* screening tests are therefore acting both to replace and reduce animal tests.

Reduction can be achieved in a number of ways:

a) some laboratories make use of an internal system to make other researchers aware when a batch of animals will be killed. Any spare organs, tissues or body fluids can be shared among researchers, thus reducing the overall numbers of animals which are needed;

b) changes in research strategy, for example, the use of non-animal methods (replacement) in the initial stages of screening for efficacy or toxicity of a compound, are now widely used within industry, and this is one reason for the decrease in the numbers of animals used in the last decade. However, research strategy is also important in other contexts. For example, it may be possible to carry out small pilot studies using few animals, which can be reviewed before committing larger numbers of animals and resources to major experiments. The statistical guidelines developed by Muller *et al* (1984), include a detailed discussion of possible research strategies;

c) proper statistical design, prior to undertaking the study, and appropriate analysis of the resulting data can make it possible to obtain results of comparable or greater precision while using fewer animals. There is evidence that poor experimental design, together with inappropriate statistical analysis of experimental results, is leading to inefficient use of animals and of scientific resources in toxicological research and testing, and in other areas of biomedical research (Altman 1982; Festing 1994, 1995); and

d) the use of inbred strains rather than genetically heterogenous outbred stocks of mice or rats can lead to a reduction in animal use. The uniformity of inbred strains means that, in general, fewer animals are needed to achieve the same level of statistical precision, or the use of the same numbers of animals will lead to better experiments with fewer incorrect results.

Investigators should recognise the potential value of obtaining statistical advice during all the stages of the planning and conduct of an experiment, in order to avoid the unnecessary use of animals.

5. REPLACEMENT

Replacement describes those methods which permit a given purpose to be achieved without conducting experiments or other scientific procedures on protected live animals. Replacement techniques have been classified as: a) relative or absolute (Russell & Burch 1959), b) direct or indirect, and c) total or partial (Balls 1997).

a) **Relative replacement:** for example, the humane killing of a vertebrate animal to provide cells, tissues and/or organs for *in vitro* studies.

Absolute replacement: animals would not need to be used at all, for example, the permanent culture of human and invertebrate cells and tissues.

b) **Direct replacement:** for example, when the skin of human volunteers or guinea-pig skin is used *in vitro* to provide information that would have been obtained from tests on the skin of live guinea-pigs.

Indirect replacement: for example, when the pyrogen test in rabbits is replaced by the *Limulus* amoebocyte lysate (LAL) test or a test based on whole human blood.

c) **Total replacement:** for example, taking a decision not to conduct an animal procedure because of lack of justification or reliability of the method, using a human volunteer instead of a guinea-pig, or testing a chemical on cells *in vitro* instead of on animals.

Partial replacement: for example, the use of non-animal methods as prescreens in toxicity testing strategies.

Many researchers who use *in vitro* or other non-animal methods do not consider these as replacements for the use of animals. Rather, the methods are used because they are the most appropriate methods for the particular area of interest, and because they can more accurately provide the information required. This is especially the case in areas such as pharmacology and biochemistry, where cellular and molecular events are studied. Other researchers use non-animal methods as adjuncts to animal experiments, to focus on specific biological functions; in many cases the overall number of animals used is reduced.

A book edited by Smith and Boyd (1991) discusses the moral issues involved in animal experimentation, and includes a chapter on the ethical considerations associated with the use of replacements.

The range of replacement methods and approaches includes the following (Balls 1994):

a) the improved storage, exchange and use of information about animal experiments already carried out, so that unnecessary repetition of animal procedures can be avoided;

b) the use of physical and chemical techniques, and of predictions based on the physical and chemical properties of molecules;

c) the use of mathematical and computer models, including: i) modelling of quantitative structure-activity relationships; ii) molecular modelling and the use of computer graphics; and iii) modelling of biochemical, physiological, pharmacological, toxicological and behavioural systems and processes;

d) the use of *in vitro* methods, including subcellular fractions, short-term maintenance of tissue slices, cell suspensions and perfused organs, and tissue culture proper (cell and organotypic culture), including human tissue culture;

e) the use of 'lower' organisms with limited sentience and/or not protected by legislation controlling animal experiments;

f) the use of the early developmental stages of vertebrates before they reach the point at which their use in experiments and other scientific procedures is regulated; and

g) human studies, including the use of human volunteers, post-marketing surveillance and epidemiology.

A great deal of effort has been placed into developing replacements to the use of animals in toxicity testing. The reasons for this include the following:

a) regulatory guidelines for the testing of many chemicals require the use of animals. There is therefore an impetus to develop methods which can replace the use of animals in regulatory toxicity testing;

b) in most cases, the strategy for conducting toxicity tests requires that some of the animals are exposed to toxic doses of a compound and this will lead to suffering. Ethical considerations have therefore led to a greater focus on the development of alternatives in toxicity testing; and

c) the acceptance of a replacement alternative by regulatory authorities is likely to lead to a greater implementation of the method into other areas of research. Therefore, the validation of replacements for the purposes of toxicity testing can be seen as an initial step in increasing the use of these methods.

As a result of the focus on alternatives in toxicity testing, many of the replacement examples in this booklet have been developed for use in toxicology. However, the majority can also be used in other areas of research.

5.1 Improved use of information

Making information available about the results of animal studies already carried out can help to avoid unnecessary repetition of animal procedures and can help with the validation of replacement methods.

Most researchers now have access to on-line databases which give up-to-date lists of published research in all areas of science (see section 6.2). Before planning an experimental procedure, a scientist should be aware of any similar work which has already taken place, with either animal or non-animal procedures, so that unnecessary repetition is avoided. Furthermore, researchers can carry out searches for potential alternatives in their area of work. Details of this approach have been outlined by Ungar (1993).

5.2 Physical and chemical techniques

Physical and chemical techniques are especially important in the study of biochemistry in general, and in related specialist fields, such as pharmacology and toxicology. In biochemistry, physical and chemical techniques can, for example, be used to study enzyme structures and mechanisms of action. In pharmacology and toxicology, physicochemical analysis can be used in predicting both the likely beneficial and harmful biological effects of chemical substances.

The use of physicochemical data, such as pH and partition coefficients, combined with structure-activity relationships, is being extensively used for the preliminary screening of chemicals.

5.3 Use of mathematical and computer modelling

5.3.1 Molecular modelling

In molecular modelling, the three-dimensional structural and electronic properties of a biological site are used to predict whether a novel molecule would interact with it, producing a biological effect. In some cases the detailed structures of such biological sites are known (for example, from X-ray crystallography, or other studies). In other cases, the nature of the site of action is inferred from the structures of molecules known to interact with it, compared with similar structures that do not (a technique known as pseudoreceptor modelling). Thus, molecular modelling is restricted to making predictions about biological sites whose structures are reasonably well understood.

An extensive literature exists on molecular modelling and many commercial systems are available, as well as public domain software. The application of powerful computational techniques has meant that computer-aided drug design is now an important non-animal screening method in pharmaceutical research and development (Richards 1994).

5.3.2 Structure-activity relationships

Molecular modelling techniques have also been applied to structure-activity prediction. Approaches concerned with the prediction of biological activity from molecular structures or physicochemical properties are known as (quantitative) structure-activity relationships ([Q]SAR), and there are currently several computer methods available which are based on a variety of techniques (Smithing & Darvas 1992; Lewis *et al* 1994). Such techniques are increasingly being used in the pharmaceutical industry to predict the effects of a slight change in the structure of a compound, before animal testing takes place.

The CASE program is an example of computer-based SAR (Klopman & Rosenkranz 1994) which links a description of chemical structure in terms of recognisable fragments (for example, rings, bonds and atoms), to biological activity. An advantage of the SAR approach is its ability to deal with quite diverse structures. As QSAR data sets are composed of physicochemical descriptors, they lend themselves to treatment by a number of statistical and mathematical techniques. The TOPKAT program makes use of a combination of substructural descriptors and physicochemical properties in order to make predictions of toxicity (Enslein *et al* 1994).

DEREK (Deductive Estimation of Risk from Existing Knowledge) is an expert system which interfaces on-screen structural information with a toxicity database so that qualitative predictions about novel structures, drawn by the toxicologist, can be made and toxicophores identified (Sanderson & Earnshaw 1991; Combes & Judson 1995). DEREK permits easy storage of, and access to, toxicity information to facilitate predictions.

In most cases, such approaches involve modelling only parts of molecules to identify specific pharmacophores, although at least one system has been developed which models entire molecules (ie COMPACT, which assesses the efficiency of the test molecule to act as a substrate for a particular isozyme of cytochrome P450). The relative advantages and disadvantages of each of these systems is discussed in a recent ECVAM/ECB (European Chemicals Bureau) workshop (Dearden *et al* 1997).

5.3.3 *PBPK modelling*

Physiologically based pharmacokinetic (PBPK) modelling predicts the disposition of xenobiotics and their metabolites by integrating three types of information: species-specific physiological parameters; partition coefficients for the chemical; and metabolic parameters. The physiological information required for developing such models is readily available from the literature. Partition coefficients can be measured by vial equilibration techniques, and metabolic constants can be estimated from experiments with isolated cells, tissue fractions, etc, or from *in vivo* metabolism studies (Krishnan & Andersen 1994; Blaauboer *et al* 1996).

Based upon this information, PBPK models can simulate the disposition of xenobiotics and their metabolites, and thus predict tissue exposure to these chemicals for various doses and species. In addition, some PBPK models include different routes of administration (for example, inhalation, oral, intravenous, dermal) of the test chemicals and could, therefore, be of value in characterising the tissue concentration achieved in each case. These models have a clear potential for refining absorption, distribution, metabolism and excretion (ADME) studies, and might also be valuable in focusing predictive *in vitro* studies of xenobiotic toxicity to concentration ranges which are likely to be found *in vivo*. A recent ECVAM workshop discussed the integrated use of QSARs, PBPK models and *in vitro* approaches for predicting toxic hazard (Barratt *et al* 1995; Blaauboer *et al* 1996).

5.3.4 *Computer programs in education*

There is a wide variety of computer programs available to study anatomy, physiology and other processes for education and training purposes (Dewhurst *et al* 1994; Dewhurst & Jenkinson 1995). Zinko *et al* (1997) provide a comprehensive guide to the use of alternative methods in education, including details of computer programs.

5.4 *In vitro* techniques

For decades, the use of *in vitro* methods has been part of the normal repertoire of biomedical research (for reviews of aspects of the use of *in vitro* methods in pharmaceutical research, see Castell and Gómez-Lechón [1997] and Rodríguez-Farré *et al* [1993]). For example, biochemists and molecular biologists work almost exclusively with isolated cells, viruses, bacteria or cell fractions; and physiologists and pharmacologists make wide use of animal and human tissues. In toxicology, *in vitro* methods act as pre-screens, so that the most toxic compounds can be screened out before less toxic compounds are tested in animal experiments. Non-animal methods in toxicology not only complement animal studies, but also permit the accomplishment of subsequent improved and more-detailed *in vitro* work.

In vitro systems have several advantages which should be considered in their development and utilisation:

a) validated *in vitro* systems can provide information in a cost-effective and time-saving manner;

b) *in vitro* systems can sometimes be used to produce data which are more reliable and more reproducible than data from animal studies. In other cases, they can be used to increase the efficiency of whole-animal studies and decrease the number of animals required;

c) *in vitro* systems are ideal for mechanistic investigations at the molecular and cellular level, as well as for target organ and target species toxicity studies;

d) human tissues can be used in *in vitro* systems. The use of human cells obviates the need for cross-species extrapolation, and thus is more relevant to the human situation; and

e) the absence of complex body systems, which might act as confounding factors, can permit studies which cannot be conducted in animals.

In vitro systems have a number of general limitations:

a) they lack the integrated systemic mechanisms of ADME;

b) they lack the complex, interactive effects of the immune, blood, endocrine and nervous systems;

c) models are not yet available for all tissues and organs, or for all toxicity endpoints;

d) it can be difficult to relate concentrations of drugs and test chemicals *in vitro* with those occurring in body fluids *in vivo*.

5.4.1 Tissue culture

The term 'tissue culture' is a generic term which covers the *in vitro* cultivation of organs, tissues, cells and embryos.

Table 1 Examples of advantages and limitations of tissue culture systems.

System	Advantages	Limitations
Organ culture	Retention of structural integrity Maintenance of cell-cell inter-relationships	Large degree of experimental variation Short-term viability Statistical sampling problems
Primary cell	Retention of several differentiated functions	Loss of tissue architecture Requires recovery period owing to damage from isolation
Continuous cell culture	Increased cell viability period Easier to maintain than primary cultures Can assume characteristics of transformed cells Can differentiate *in vitro* to assume new phenotypes	Loss of tissue architecture Loss of differentiated organ functions Can assume characteristics of transformed cells Significant loss of metabolising capacity

5.4.2 *Organ culture*

Organ culture refers to a three-dimensional culture of tissue retaining some or all of the histological features of the tissue and preservation of its architecture, usually by culturing the tissue at the liquid-gas interface on a grid or gel. Organ cultures cannot be propagated and experiments in organ cultures generally involve a large degree of experimental variation between replicates, making organ cultures less suitable than cell cultures for quantitative determinations. The production of organ cultures requires fresh tissue from the relevant organ, although one organ can provide material for several cultures. Organ cultures can be used to study the pharmacokinetics of such processes as ADME. They are also used to study pharmacodynamics. The functions studied include oxygen consumption, glucose uptake and release, pyruvate release, lactate release, nitrogenous excretion and cell proliferation.

The culture of tissue explants (for example, liver slices), has received extensive use in recent years as a means of maintaining differentiated tissue function in an *in vitro* model. The use of tissue slices has various advantages over isolated cells, such as easier and faster preparation, the presence of different cell types, and the maintenance of cell-cell and cell-matrix interactions. A comprehensive overview of the use of tissue slices for pharmacotoxicology studies can be found in Bach *et al* (1996).

Whole skin or keratomed skin slices from many species, including human skin, have been used in a variety of assays for *in vitro* skin irritancy and corrosivity prediction (Council of European Communities 1990). Cultures of isolated skin are widely used for predicting skin absorption characteristics (Bronaugh 1995). In addition, complex human skin models have been developed in recent years and some models are available commercially, such as EpiDerm™. Test protocols have been developed for use with these human skin models, which enable topically applied aqueous and non-aqueous test materials to be screened for their skin irritancy potentials (Frazier 1991).

The study of isolated organs and tissues has produced a number of procedures which have been proposed as alternatives to the Draize eye test. These include using the hen's egg test (chorioallantoic membrane [HET-CAM] assay; Gilleron *et al* 1996), the isolated rabbit eye and bovine cornea (Atkinson *et al* 1992; Sivak *et al* 1994). However, recent validation studies have shown that more *in vitro* test development is required if the Draize eye test is to be replaced by a non-animal testing strategy (Balls *et al* 1995b; Spielmann *et al* 1996; Brantom *et al* 1997).

5.4.3 *Primary cell culture*

A primary cell culture involves the isolation of cells by the disruption of the tissue, often with proteolytic enzymes. The major advantages of primary cultures are the retention of the capacity for biotransformation and tissue-specific functions. One limitation of primary cultures is the necessity to isolate cells for each experiment. This may result in the loss or damage of specific membrane receptors, damage to the integrity of the membrane and loss of cellular products. During the interval necessary to establish monolayer cultures, damage is often repaired. Fry *et al* (1995) have reviewed the advantages and disadvantages associated with the use of freshly-isolated hepatocytes and cultured hepatocytes.

Primary cultures have a limited life-span, and changes in metabolism and tissue-specific functions will occur with time in culture. An effective means of storing monolayers of cells would be valuable for optimising the use of animal and human tissues, and work is being conducted to improve the storage of monolayers.

Cell culture techniques are commonly used for monoclonal antibody production (Schade *et al* 1996; Marx *et al* 1997), virus vaccine production (Hendriksen *et al* 1994), vaccine potency testing (Hendriksen & van der Gun 1995), screening for cytotoxic effects, and studying the function and make-up of cells.

5.4.4 Cell lines

The use of cell lines provides a number of advantages for the study of many biological phenomena, including toxicity. A wide variety of cell types can be used, including human cells and those from specific tissues, for investigations of target organ specificity.

Most cell lines can be stored at ultra low temperatures and they either have a finite life span, or they are capable of an unlimited number of population doublings (continuous cell line). However, the differentiated functions may be altered in cell lines depending on the culture conditions, and the metabolic capacity of the cells may decrease depending on the culture conditions. Although continuous cell lines offer the advantage of indefinite growth and ease of manipulation, finite cell lines are usually chromosomally abnormal and display *in vitro* ageing (Barile 1997).

A wide range of endpoints and a comprehensive selection of assays are available with cell cultures, with flexible protocols regarding dosage and exposure. Such cultures enable the study of short-term and long-term (repeat or continuous) exposure, with the possibility of assessing recovery outside the whole organism. Isolated and cultured cells offer many advantages both in the study of mechanisms of cellular toxicity by chemical agents and for *in vitro* bioassays. These systems have the advantage of ease of use and reproducibility, but frequently lack (or express in greatly reduced amounts) the differentiated functions of the *in vivo* tissue from which they were isolated. Also, established animal cell lines, such as Chinese hamster ovary (CHO) and mouse lymphoma L5178Y, suffer from interlaboratory variation, especially in basic properties like chromosome number. This is because they have been maintained for many years in culture collections.

There have been several efforts to minimise the problems of dedifferentiation, senescence and instability in cell cultures, with varying success. One approach has been to generate immortalised cell lines by introducing viral oncogenes into primary cells (MacDonald 1993). In principle, any cell type can be used. Such cell lines combine the advantages of primary and transformed cells in that they express tissue-specific characteristics and are capable of extended, often indefinite, growth *in vitro*. They can be used to isolate novel, functional cell lines, to replace the need to use primary cells, such as freshly obtained hepatocytes. Oncogenes like SV40 large T, as well as polyoma virus large T, and *myc* have been used with calcium phosphate or electroporation treatment of the cells to achieve cellular uptake. Examples of cells which have been immortalised include rabbit kidney, mouse macrophages, rat liver and human lymphocytes (MacDonald *et al* 1994), although their usefulness is a matter of controversy at present.

A recent development in cell culture techniques concerns the increasing use of membrane filter inserts. These involve the culturing of cells on the surfaces of insert membranes which fit inside larger compartments, which can contain other cells or medium and/or test material (Clothier *et al* 1995). Such two-compartment systems facilitate the study of phenomena like transepithelial transport and penetration, production of metabolites and metabolic activation and detoxification by co-cultures (for example, having hepatocytes growing in the inserts and indicator cells in the other compartment). In addition, cell recovery can be studied, since target cells can be separated from test material easily at any time after initiation of exposure.

The utilisation of molecular biological techniques is contributing to the improvement of cell lines in many other ways; for example, the introduction of shuttle vectors as target DNA for genotoxicity studies (Yagi *et al* 1994), and other genes producing specific intracellular recognition targets for pharmaceutical development studies. In addition, many different strains of indicator cells, expressing cloned cytochrome P-450 isozymes, have been generated (Combes 1992; Crespi 1995). Thus far, this approach has been more successful using mammalian cells, where a variety of cell lines has been developed, possessing several different isozymes and other enzymes, involved in the metabolism of foreign compounds by mixed function oxidases and other enzymes. Examples are human lymphoblastoid cells (AHH-1; MCL-5) and rodent fibroblasts, in which mutation at the HGPRT gene is measured.

5.4.5 *Subcellular fractions*

Fractionated organelles and membranes prepared from defined cell types can be used for specific cell-free investigations. These *in vitro* systems, which are usually prepared by differential centrifugation, include nuclei, mitochondria, lysosomes and membrane vesicles.

5.5 Use of lower organisms

Wherever possible, the least sentient species should be used in research. The term lower organisms includes invertebrates (with the exception of the common octopus, *Octopus vulgaris*), plants, fungi, bacteria and viruses. These groups of organisms, which are considered to be non-sentient, are not covered by the *Animals (Scientific Procedures) Act 1986*, and their use in experiments can reduce and replace the number of higher organisms used. Although invertebrate assays can only give limited information and cannot act as complete replacements currently in toxicology, they are useful as pre screen systems, especially for agrochemicals and environmental pollutants.

The Ames test has been validated for regulatory toxicology purposes as a screen for genotoxic chemicals. This assay utilises *Salmonella* bacteria to detect chemical mutagenesis (Mitchell & Combes 1997).

The fruit fly (*Drosophila spp.*) and the nematode (*Caenorhabditis elegans*) are widely used in genetic research and in fundamental studies in cell biology. Studies with Drosophila provided the first evidence that X-rays are mutagenic, and systems have been developed for using them as pre-screens to test for chemical mutagens and carcinogens (Mitchell & Combes 1994). Substantial progress has been made in the complete sequencing of the genome of *C. elegans* and it is becoming clear that many genes found in both these organisms are conserved in humans. *C. elegans* has already made important contributions to fundamental studies in neurophysiology and behaviour, developmental biology, including mechanisms of programmed cell death (apoptosis), and genetics. There is enormous scope for using these species in fundamental studies in many disciplines, including toxicology.

A simple test for detecting teratogenic chemicals employs the use of the coelenterate *Hydra attenuata*. The assay, which has been extensively developed and is commercially available, is based on the ability of *Hydra* to rapidly regenerate from dissociated polyps, which are used as artificial embryos (Davies & Freeman 1995).

The LAL assay employs the use of the horseshoe crab (*Limulus polyphemus*) for the detection of endotoxins, which can induce many clinical effects in humans, ranging from fever to systemic toxicity (Flint 1994).

5.6 Use of early developmental stages of vertebrates

The use of the early developmental stages of vertebrates before they become protected animals is considered a replacement. In the case of 1986 Act, the early stage is considered to be before half-way through gestation (mammals) or incubation (birds and reptiles), or the stage at which independent feeding occurs (amphibians and fish).

laevis, avian embryos and mammalian whole-embryo cultures has been discussed elsewhere (Brown *et al* 1995; Piersma *et al* 1996). These authors conclude that tests based on mammalian whole-embryo culture systems are well developed for *in vitro* methods for the detection of reproductive toxicity and teratogenicity, and for elucidating mechanisms of teratogenesis.

5.7 Use of human tissues and human volunteers

The use of human tissues avoids the problem of inter-specific extrapolation from animals to humans, and provides more relevant mechanistic information. Using human cells in culture can contribute significantly to the process of drug development, particularly with regard to metabolic studies (Wrighton *et al* 1993). Human tissues have been used to establish organotypic culture models (for example, skin equivalents). However, as problems with supply of fresh tissue exist, immortalised human cell lines have been established, mainly by DNA virus oncogene transfection. For example, the spontaneously immortalised keratinocyte cell line, HaCaT, exhibits a normal differentiation pattern under optimised *in vitro* conditions and maintains a nearly normal expression of metabolic enzymes (Fusenig *et al* 1991).

There are, however, both logistical and ethical problems with obtaining human tissue samples. Logistical problems include safety issues associated with the necessity to screen for HIV and hepatitis, to obtain detailed histories of donors relating to genetic background, drug intake and details of any other exposures, as well as the requirement for a system to keep track of tissues obtained from each donor, in case of future problems. The legal situation varies according to country. In the UK, the use of human tissues is regulated such that prior informed consent for the use of cells, tissues and transplant organs has to be obtained from donors/relatives and from local research ethics committees. There are also potential ownership problems with regard to the data and any commercial/legal implications arising therefrom. The use of foetal tissue needs careful consideration, since it has its own complex ethical aspects.

Sources of human tissue include the International Institute for the Advancement of Medicine (IIAM) and the Stephen Kirby Skin Bank and Research Unit (see section 5.8 for details).

Human volunteer studies are increasingly being used in dermal toxicology and the development of methods for human monitoring is a very active part of genetic toxicology. In the field of human monitoring, the development of extremely sensitive immunological methods looks very promising. Furthermore, the development of safe, non-invasive and highly sensitive techniques to monitor the effects of chemicals, such as magnetic resonance imaging (MRI) and magnetic resonance spectroscopy (MRS) are facilitating more human studies (Brauer 1993). The high quality images obtained with MRI can provide information on the distribution of water and electrolytes, fat deposition and the development of oedema in a tissue. MRS permits serial measurement of xenobiotics, metabolites, and endogenous compounds in cells and tissue, as well as in samples of body fluids. Further advances in MRS technology and increased computing power for data analysis raise the possibility of MRS being used in future to investigate even more-complex structures, for example, receptor interactions and perturbations in cellular membranes.

5.8 Sources of tissues

Cell lines are available from the European Collection of Animal Cell Cultures (ECACC) Porton Down, UK;

Tel. 01980 612512; Fax. 01980 611315; E-mail. ecacc@ecacc.demon.co.uk.

Human cell systems are available by contacting TCS Biologicals Ltd, Botolph Claydon, Buckingham, MK18 2LR;

Tel. 01296 714071; Fax. 01296 715752; E-mail. sales@tcsgroup.co.uk.

The German Collection of Human and Animal Cell Cultures distributes a wide range of cell lines. Their Web site is located at http://www.biotech.ist.unige.it/cldb/coll121.html.

The International Institute for the Advancement of Medicine (IIAM) in Leicester supplies human non-transplantable tissues, provided that a research proposal is approved. For information, contact IIAM, Medical Sciences Building, University Road, Leicester, LE1 9HN;

Tel. 0116 252 3047; Fax. 0116 252 5680.

Skin cells can be obtained from the Stephen Kirby Skin Bank and Research Unit, Queen Mary's University Hospital, Roehampton, London, SW15 5PN;

Tel. 0181 780 1145; Fax. 0181 780 1765.

5.9 Conclusion

Advances in computer and *in vitro* technology have permitted a larger variety of modelling systems and tissue cultures. The rapid expansion of knowledge in the fields of molecular biology and immunology have provided techniques that enable the detection of very early events at a cellular level, where previously it was necessary to observe large-scale manifestations in the whole animal.

In vitro methods can be used to answer certain questions which cannot be addressed with *in vivo* models. Furthermore, the use of human tissues is more relevant to the human situation than the use of animals. In other cases, the use of alternatives in the early stages of research strategy to screen for efficacy or safety can reduce the numbers of animals which are later required. Other advantages of *in vitro* methods include their low cost, high reproducibility, ease of use and highlights the advantages and limitations of non-animal systems, to aid scientists when choosing the most suitable methods for investigations in their areas of interest.

6. SOURCES OF INFORMATION ON THE THREE Rs

6.1 Literature and organisations

Information on alternatives to the use of animals can be obtained from a variety of services. Addresses of the information sources highlighted in bold are listed at the end of this section.

For enquiries concerning the 3Rs, contact the following organisations:

ECVAM, FRAME, The Dr Hadwen Trust for Humane Research, The Humane Research Trust, UFAW.

A number of scientific journals deal specifically with non-animal techniques and methods for improvements in animal welfare: *ATLA (Alternatives To Laboratory Animals)*; *In Vitro Toxicology*; *Toxicology in Vitro*; *Laboratory Animals*; and *Animal Welfare*.

ATLA also publishes the reports and recommendations of ECVAM workshops; reports are available by writing to Dr Julia Fentem at **ECVAM**. At the time of writing, the following workshop reports are available:

1. The practical applicability of hepatocyte cultures in routine testing
2. *In vitro* phototoxicity testing
3. *In vitro* neurotoxicity testing
4. Alternatives to animal testing in the quality control of immunobiologicals: current status and future prospects
5. Practical aspects of the validation of toxicity test procedures
6. A prevalidation study on *in vitro* skin corrosivity testing
7. Development and validation of non-animal tests and testing strategies: the identification of a coordinated response to the challenge and the opportunity presented by the Sixth Amendment to the Cosmetics Directive (*76/768/EEC*)
8. The integrated use of alternative approaches for predicting toxic hazard
9. Safety and efficacy testing of hormones and related products
10. Nephrotoxicity testing *in vitro*
11. The 3Rs: the way forward
12. Screening chemicals for reproductive toxicity: the current alternatives
13. Methods for assessing percutaneous absorption
14. The use of *in vitro* systems for evaluating haematotoxicity
15. The use of biokinetics and *in vitro* methods in toxicological risk evaluation
16. Acute toxicity testing *in vitro* and the classification and labelling of chemicals
17. Alternatives to the animal testing of medical devices

18. *In vitro* tests for respiratory toxicity
19. Alternative methods for skin sensitisation testing
20. The use of tissue slices for pharmacotoxicology studies
21. The production of avian (egg yolk) antibodies: IgY
22. Pharmacokinetics in early drug research
23. Monoclonal antibody production
24. The development and validation of expert systems for predicting toxicity
25. Current status and future developments of databases on alternative methods.

6.2 On-line services and other databases

Access to the large computerised biosciences databases, such as Medline, Embase, Biosis and Agricola, is through the services of a host, for example, Dialog or DataStar (both from Knight-Ridder and soon to be combined). The **PREX** on-line information service based at Utrecht University is a host service that offers a unique combination of general and more-specialised databases, with an emphasis on information concerned with animal welfare, laboratory animal science and alternatives to animal use. Access to PREX is via the Internet, by using telnet with ftp and rcp-protocol for downloading records, or via a direct modem-to-modem connection. Some of the databases currently available through PREX include: Medline (biomedicine), CSA Life Sciences (twenty life science disciplines), CAB-International (veterinary science and animal welfare), Agricola (agricultural and life sciences, including animal welfare), AT Alternatives (literature citations on alternatives to animals, produced by the Akademie für Tierschutz, Germany), Biomedical Dissertations (from Dutch universities), Current Citation (the most frequently requested journals for document delivery by the British Library), **NORINA** (index of audiovisual and multimedia resources to replace the use of animals in education), and Serline (index of biomedical journals), as well as a number of smaller veterinary and laboratory animal science databases. Although the user interface is not as user-friendly as, for example, the Windows-based interfaces available from Knight-Ridder, PREX is a very cost-effective service, which permits 120 hours of search time and unlimited downloading of selected records for a single annual subscription of approximately £150-250, depending on the number of core databases required.

Retrieval problems can occur in the very large databases. For example, the Medline index term 'animal testing alternatives' is commonly used only in relation to articles which discuss the need for alternatives. No attempt is made to assess whether an experimental paper describes a potential alternative to the use of animals. Likewise, the check tag '*in vitro*' does not appear to be used in a consistent manner by Medline indexers. Therefore, it is necessary to have both a specialised knowledge of the field of study and a detailed knowledge of the database structure and indexing system, in order to construct the most effective search profile for a search on potential alternatives. Since no one database can provide exhaustive coverage of any field, the same search should be carried out in a number of databases, each of which will require the construction of a separate search profile to take into account the different approaches to indexing.

Two bibliographical services are available which derive from large on-line databases: The **Animal Welfare Information Centre** (AWIC) of the US Department of Agriculture publish a series of non-exhaustive current-awareness *Quick Bibliographies* consisting of citations downloaded from their Agricola database. Some of the selected topics relate to the use of animals and alternatives. Likewise, the **National Library of Medicine** produces a selective annotated bibliography on alternatives based on regular searches of their databases, including Medline.

INVITTOX protocols are a constantly increasing collection of detailed methodological descriptions of *in vitro* experimental and test systems of potential interest to toxicologists. The service was established by FRAME and is in the process of being integrated into a wider system of databases on methods, expertise and validation of alternatives that is being established at ECVAM. Each protocol aims to provide sufficient detail to allow a scientist to perform the methods described. A total of 114 protocols are currently available, but only in printed form. The intention is eventually to make all protocols available in their full form on the World Wide Web as part of the ECVAM Scientific Information Service.

Finally, the potential of Internet mailing lists and newsgroups as a source of information should not be underestimated. Both these structures provide a forum in which to ask questions of colleagues working in the same field. The Electronic Zoo is a list of these and other Internet resources that relate to laboratory animals and their welfare. It can be accessed from the **NetVet** Web site which features other resources such as electronic publications from AWIC.

AWIC

Animal Welfare Information Center, National Agricultural Library, 10301 Baltimore Avenue, 5th Fl., Beltsville, MD 20705, USA. Fax: 00 1 301 5047125. AWIC can also be accessed via the NetVet Web page (see below) and through its own page at http://www.nalusda.gov/awic/awic.htm

The Dr Hadwen Trust for Humane Research

22 Bancroft, Hitchin, Herts, SG5 1JW. Fax: 01462 436819

ECVAM

JRC Environment Institute, 21020 Ispra (Va), Italy. Fax: 00 39 332 785336. E-mail: julia.fentem@jrc.it

ECVAM Scientific Information Service

details from Annett Janusch, ECVAM. E-mail: annett.janusch@jrc.it

FRAME

Russell & Burch House, 96-98 North Sherwood Street, Nottingham, NG1 4EE. Fax: 0115 9503570.

E-mail: frame@frame-uk.demon.co.uk
Web site at
http://www.frame-uk.demon.co.uk/

The Humane Research Trust

Brook House, 29 Bramhall Lane South, Bramhall, Cheshire, SK7 2DN. Fax: 0161 439 3712

INVITTOX

details from Krys Bottrill, FRAME. E-mail: invittox@frame-uk.demon.co.uk

National Library of Medicine

Alternatives to the Use of Live Vertebrates in Biomedical Research and Testing - an Annotated Bibliography. Further information on this publication can be obtained from NLM Specialized Information Services, Office of Hazardous Substances Information, Building 38A, Room 55-516, 8600 Rockville Pike, Bethesda, MD 20894, USA. E-mail: vera_hudson@ccmail.nlm.nih.gov. The bibliography is at present on the NLM gopher service, which can be accessed from the NLM home page at http://www.nlm.nih.gov

NetVet

Web site at http://netvet.wustl.edu

NORINA

Web site at
http://oslovet.veths.no/NORINA/default.html

PREX

PREX, Utrecht University, P.O.B. 80166, NL 3508 TD Utrecht, The Netherlands. Fax. 00 31 30 2536747. E-mail: prex@pdk.dgk.ruu.nl

UFAW

The Old School, Brewhouse Hill, Wheathampstead, Herts, AL4 8AN. Fax: +44 (0) 1582 831414.

E-mail: ufaw@ufaw.org.uk
Web site at
http://www.users.dircon.co.uk~ufaw3/

7. REFERENCES

Altman D G 1982 Statistics in medical journals. *Statistics in Medicine 1*: 59-71

Anon 1986a *Animals (Scientific Procedures) Act 1986*. 24pp. HMSO: London, UK

Anon 1986b Council Directive of 24 November 1986 on the approximation of laws regulations and administrative provisions of the Member States regarding the protection of animals used for experimental and other scientific purposes. *Official Journal of the European Communities 29*: *(L358):* 1-29

Anon 1986c *European Convention for the Protection of Vertebrate Animals Used for Experimental and Other Scientific Purposes*. 51pp. Council of Europe: Strasbourg

Anon 1996 *Report of the OECD Workshop on Harmonisation of Validation and Acceptance Criteria for Alternative Toxicological Test Methods*. OECD Paris, France

Atkinson K A, Fentem J H, Clothier R H and Balls M 1992 Alternatives to ocular irritation testing in animals. *Lens and Eye Toxicity Research 9*: 247-258

Bach P H, Vickers A E M, Fisher R, Baumann A, Brittebo E, Carlile D J, Koster H J, Lake B G, Salmon F, Sawyer T W and Skibinski G 1996 The use of tissue slices for pharmacotoxicology studies. The report and recommendations of ECVAM workshop 20. *ATLA 24*: 893-923

Balls M 1994 Replacement of animal procedures: alternatives in research education and testing. *Laboratory Animals 28*: 193-211

Balls M 1997 Non-animal alternatives and reduction in laboratory animal use. *ATLA 25*: 215-218

Balls M, Blaauboer B J, Bruner L, Combes R D, Eckwall B, Fentem J H, Fielder R, Guillouzo A, Lewis R, Lovell D, Reinhardt C, Repetto G, Sladowski D, Spielmann H and Zucco F 1995a Practical aspects of the validation of toxicity test procedures. The report and recommendations of ECVAM Workshop 5. *ATLA 23*: 129-147

Balls M, Botham P S, Bruner L H and Spielmann H 1995b The EC/HO international validation study on alternatives to the Draize eye irritation test. *Toxicology in Vitro 9*: 871-929

Balls M, Goldberg A M, Fentem J H, Broadhead C L, Burch R L, Festing M F W, Frazier J M, Hendriksen C F M, Jennings M, van der Kamp M D O, Morton D B, Rowan A N, Russell C, Russell W M S, Spielmann H, Stephens M L, Stokes W S, Straughan D W, Yager J D, Zurlo J and van Zutphen B F M 1995c The Three Rs: the way forward. The report and recommendations of ECVAM workshop 11. *ATLA 23*: 838-866

Balls M and Fentem J H 1997 Progress toward the validation of alternative tests. *ATLA 25*: 33-43

Barile F A 1997 Continuous cell lines as a model for drug toxicity assessment. In: J V Castell and M J Gómez-Lechón (eds) *In Vitro Methods in Pharmaceutical Research* pp 33-54. Academic Press Ltd: London, UK

Barratt M D, Castell J V, Chamberlain M, Combes R D, Dearden J C, Fentem J H, Gerner I, Giuliani A, Gray T J B, Livingstone D J, Provan W M, Rutten F A J J L, Verhaar H J M and Zbinden P 1995 The integrated use of alternative approaches for predicting toxic hazard. The report and recommendations of ECVAM workshop 8. *ATLA 23*: 410-429

Blaauboer B J, Bayliss M K, Castell J V, Evelo C T A, Frazier J M, Groen K, Galden M, Guillouzo A, Hissink A M, Houston J B, Johanson G, de Jongh J, Kedderis G L, Reinhardt C A, van de Sandt J J M and Semino G 1996 The use of biokinetics and *in vitro* methods in toxicological risk evaluation. The report and recommendations of ECVAM workshop 15. *ATLA 24*: 473-497

Brantom P G, Bruner L H, Chamberlain M, De Silva O, Dupuis J, Earl L K, Lovell D P, Pape W J W, Uttley M, Bagley D M, Baker R W, Bracher M, Courtellemont P, Declercq L, Freeman S, Steiling W, Walker A P, Carr G J, Dami N, Thomas G, Harbell J, Jones P A, Pfannenbecker U, Southee J A, Tcheng M, Argembeaux H, Castelli D, Clothier R, Esdaile D J, Itigaki H, Jung K, Kasai Y, Kojima H, Kristen U, Larnicol M, Lewis R W, Marenus K, Moreno O, Peterson A, Rasmussen E S, Robles C and Stern M 1997 A summary report of the COLIPA international validation study on alternatives to the Draize eye irritation test. *Toxicology in Vitro 11*: 141-179

Brauer M 1993 Magnetic resonance imaging and spectroscopy: new non-invasive *in vivo* approaches in toxicology research. *ATLA 21*: 411-426

Bronaugh R L 1995 Methods for *in vitro* percutaneous absorption. *Toxicology Methods 5*: 265-273

Brown N A, Spielmann H, Bechter R, Flint O P, Freeman S J, Jelinek R J, Koch E, Nau H, Newall D R, Palmer A K, Renault J-Y, Repetto M F, Vogel R and Viger R 1995 Screening chemicals for reproductive toxicity: the current alternatives. The report and recommendations of ECVAM/ETS workshop (ECVAM workshop 12, *ATLA 23*: 868-882

Castell J V and Gómez-Lechón M J 1997 *In Vitro Methods in Pharmaceutical Research* 467pp. Academic Press: London, UK

Clothier R H, Atkinson K A, Garle M J, Ward R K and Willshaw A 1995 The development and evaluation of *in vitro* tests by the FRAME Alternatives Laboratory. *ATLA 23*: 75-90

Combes R D 1992 The *in vivo* relevance of *in vitro* genotoxicity assays incorporating enzyme activation systems. In: Gibson G G (ed) *Progress in Drug Metabolism Volume 13* pp 295-321. Taylor and Francis Ltd: London, UK

Combes R D and Judson P 1995 The use of artificial intelligence systems for predicting toxicity. *Pesticide Science 45*: 179-194

Council of European Communities 1990 *Skin irritation. In: ECETOC Monograph No. 15* CEC: Brussels

Crespi C 1995 Use of genetically engineered cells for genetic toxicology testing. In: Phillips DH and Venitt S (eds) *Environmental Mutagenesis* pp 233-260. Bios Scientific Publishers: Oxford, UK

Davies W J and Freeman S J 1995 *In vitro* toxicity testing protocols. In: Hare SO and Atterwill C (eds) *Methods in Molecular Biology* pp 321-326. Humana Press Inc: New Jersey, USA

Dearden J C, Barratt M D, Benigni R, Bristol D W, Combes R D, Cronin M T D, Judson P M, Payne M P, Richard A M, Tichy M, Worth A P and Yourick J J 1997 The development and validation of expert systems for predicting toxicity. The report and recommendations of an ECVAM/ECB workshop (ECVAM workshop 24, *ATLA 25*: 223-252

Dewhurst D, Hughes I and Ullyott R 1994 The Finkleman Preparation: a computer simulation for teaching undergraduate students of pharmacology. *ATLA 22*: 474-482

Dewhurt D and Jenkinson L 1995 The impact of computer-based alternatives on the use of animals in undergraduate teaching: a pilot study. *ATLA 23*: 521-531

Enslein K, Gombar V K and Blake B W 1994 Use of SAR in computer-assisted prediction of carcinogenicity and mutagenicity of chemicals by the TOPKAT program. *Mutation Research 305*: 47-62

Festing M F W 1994 Reduction of animal use: experimental design and quality of experiments. *Laboratory Animals 28*: 212-221

Festing M F W 1995 Reduction in animal use 35 years after Russell & Burch's *Principles of Humane Experimental Technique*. *ATLA 23*: 51-60

Flecknell P A 1994 Refinement of animal use: assessment and alleviation of pain and distress. *Laboratory Animals 28*: 222-231

Flint O 1994 A timetable for replacing reducing and refining animal use with the help of *in vitro* tests: the Limulus amoebocyte lysate test (LAL) as an example. In: Reinhardt CA (ed) *Alternatives to Animal Testing. New Ways in the Biomedical Sciences Trends and Progress* pp 27-43. VCH: Weinheim, Germany

Frazier J M 1991 Alternatives to acute ocular and dermal toxicity tests in animals. In: Marzulli FN and Maibach HI (eds) *Dermatotoxicology 4th edn* pp 817-833. Hemisphere Publishing Corporation: New York, USA

Fry J R, Hammond A H, Atmaca M, Dhanjal P and Wilkinson D J 1995 Toxicity testing with hepatocytes: some methodological aspects. *ATLA 23*: 91-96

Fusenig N E, Boukamp P, Breitkreutz D and Hulsen A 1991 Altered regulation of growth and differentiation at different stages of transformation of human skin keratinocytes. In: Rhim JS and Dritschilo A (eds) *Neoplastic Transformation in Human Cell Culture* pp 235-250. Humana Press Inc: New Jersey, USA

Gilleron L, Coecke S, Sysmans M, Hansen E, Van Oproy S, Marzin D, Van Cauteren H and Vanparys Ph 1996 Evaluation of modified HET-CAM assay as a screening test for eye irritancy. *Toxicology in Vitro 10*: 431-446

Hendriksen C F M, Garthoff B, Aggerbeck H, Bruckner L, Castle P, Cussler K, Dobbelaer R, van de Donk H, van der Gun J, Lefrancois S, Milstien J, Minor P D, Hougeot H, Rombaut B, Ronneberger H D, Spieser J-M, Stolp R, Straughan D W, Tollis M and Zigtermans C 1994 Alternatives to animal testing in the quality control of immunobiologicals: current status and future prospects. The report and recommendations of ECVAM workshop 4. *ATLA 22*: 420-434

Hendriksen C and van der Gun J 1995 Animal models and alternatives in the quality control of vaccines: are *in vitro* methods or *in vivo* methods the scientific equivalent of the emperor's new clothes? *ATLA 23*: 61-74

Klopman G and Rosenkranz H S 1994 Approaches to SAR in carcinogenesis and mutagenesis: prediction of carcinogenicity/mutagenicity using MULTI-CASE. *Mutation Research 305*: 33-46

Krishnan K and Andersen M E 1994 Physiologically-based pharmacokinetic modelling in toxicology. In: Hayes AW (ed) *Principles and Methods of Toxicology 3rd edn* pp 149-188. Raven Press Ltd: New York, USA

Lewis D F V, Moereels H, Lake B G, Ioannides C and Parke D.V 1994 Molecular modelling of enzymes and receptors involved in carcinogenesis: QSARs and COMPACT-3D. *Drug Metabolism Reviews 26*: 261-285

MacDonald C 1993 Development of new animal cell lines by oncogene immortalisation. *The Genetic Engineer and Biotechnologist 13*: 121-124

MacDonald C, Vass M, Willett B, Scott A and Grant H 1994 Expression of liver functions in immortalised rat hepatocyte cell lines. *Human and Experimental Toxicology 13*: 439-444

Marx U, Embleton M J, Fischer R, Gruber F P, Hansson U, Heuer J, de Leeuw W A, Logtenberg T, Merz W, Portetelle D, Romette J-L and Straughan D W 1997 Monoclonal antibody production. The report and recommendations of ECVAM workshop 23. *ATLA 25*: 121-137

Mitchell I de G and Combes R D 1984 Mutation tests with the fruit fly *Drosophila melanogaster*. In: Venitt S and Parry JM (eds) *Mutagenicity Testing: A Practical Approach* pp 149-185. IRL Press: Oxford, UK

Mitchell I de G and Combes R D 1997 *In Vitro* genotoxicity and cell transformation assessment. In: Castell JV and Gomez-Lechon MJ (eds) *In vitro Methods in Pharmaceutical Research* pp 317-352. Academic Press: London, UK

Morton D B 1995 Advances in refinement in animal experimentation over the past 25 years. *ATLA 23*: 812-823

Morton D B, Abbot A, Close B S, Ewbank R, Gask D, Heath M, Mattic S, Poole T, Seamer J, Southee J, Thompson A, Trussell B, West C and Jennings M 1993a First report of the BVAAWF/FRAME/RSPCA/UFAW Joint Working Group on Refinement: removal of blood from laboratory mammals and birds. *Laboratory Animals 27*: 1-22

Morton D B, Jennings M, Batchelor G R, Bell D, Birke L, Davies K, Eveleigh J R, Gunn D, Heath M, Howard B, Koder P, Phillips J, Poole T, Sainsbury T, Sales G D, Smith D J A, Stauffacher M and Turner R J 1993b Second report of the BVAAWF/FRAME/RSPCA/UFAW Joint Working Group on Refinement: refinements in rabbit husbandry. *Laboratory Animals 27*: 301-329

Muller K E, Barton C N and Benignus V A 1984 Recommendations for appropriate statistical practice in toxicological experiments. *Neurotoxicology 5*: 113-126

Piersma A H, Bechter R, Krafft N, Schmid B P, Stadler J, Verhoef A, Verseil C and Ziljstra J 1996 An interlaboratory evaluation of five pairs of teratogens and non-teratogens in post-implantation rat embryo culture. *ATLA 24*: 201-209

Poole T 1997 Happy animals make good science. *Laboratory Animals 31*: 116-124

Richards W G 1994 Computer-aided drug design. *Pure and Applied Chemistry 66*: 1589-1596

Rodríguez-Farré E, Roberfroid M and Fracchia G N 1993 Research and development of *in vitro* pharmacotoxicology: a European perspective. *ATLA 21*: 285-293

Routledge E J and Sumpter J P 1996 Estrogenic activity of surfactants and some of their degradation products assessed by a recombinant yeast screen. *Environmental Toxicology and Chemistry 15*: 241-248

Russell W M S and Burch R L 1959 *The Principles of Humane Experimental Technique* 238pp. Methuen: London, UK

Sanderson D M and Earnshaw C G 1991 Computer prediction of possible toxic action from chemical structure: the DEREK system. *Human and Experimental Toxicology 10*: 261-273

Schade R, Staak C, Hendriksen C, Erhard M, Hugl H, Koch G, Larsson A, Pollmann W, van Regenmortel M, Rijke E, Spielmann H, Steinbusch H and Straughan D 1996 The production of avian (egg yolk) antibodies: IgY. The report and recommendations of ECVAM workshop 21. *ATLA 24*: 925-934

Sivak J G, Herbert K L and Segal L 1994 Ocular lens organ culture as a measure of ocular irritancy: the effects of surfactants. *Toxicology Methods 4*: 56-65

Smith J A and Boyd K M 1991 *Lives in the Balance: The Ethics of Using Animals in Biomedical Research*, 352pp. Oxford University Press: Oxford, UK

Smithing M P and Darvas F 1992 Hazardexpert: an expert system for predicting chemical toxicity. In: Finlay JW, Robinson SF and Armstrong DJ (eds) *Food Safety Assessment* pp 191-200. American Chemical Society: Washington, USA

Spielmann H, Liebsch M, Kalweit S, Moldenhauer F, Wirnsberger T, HolzhÅtter H-G, Schneider B, Glaser S, Gerner I, Pape W J W, Kreiling R, Krauser K, Miltenburger H G, Steiling W, Luepke N P, MÅller N, Kreuzer H, MÅrmann P, Spengler J, Bertram-Neis E, Siegemund B and Wiebel F J 1996 Results of a validation study in Germany on two *in vitro* alternatives to the Draize eye irritation test the HET-CAM test and the 3T3 NRU cytotoxicity test. *ATLA 24*: 741-859

Ungar K 1993 *INVITTOX*: an attempt to solve some information problems of *in vitro* toxicology. *Humane Innovations and Alternatives 7*: 540-543

Wrighton **S A, Vandenbranden M, Stevens J C, Shipley L A, Ring B J, Rettie A E and Cashman J R** 1993 *In vitro* methods for assessing human hepatic drug metabolism: their use in drug development. *Drug Metabolism Reviews 25*: 453-484

Yagi T, Sato M, Nishigori C and Takebe H 1994 Similarity in the molecular profile of mutations induced by UV light in shuttle vector plasmids propagated in mouse and human cells. *Mutagenesis 9*: 73-77

Zinko U, Jukes N and Gericke C 1997 *From Guinea Pig to Computer Mouse* 229pp. EuroNICHE.

APPENDIX: A strategy for the assessment of the potential for implementation of the Three Rs into a research protocol

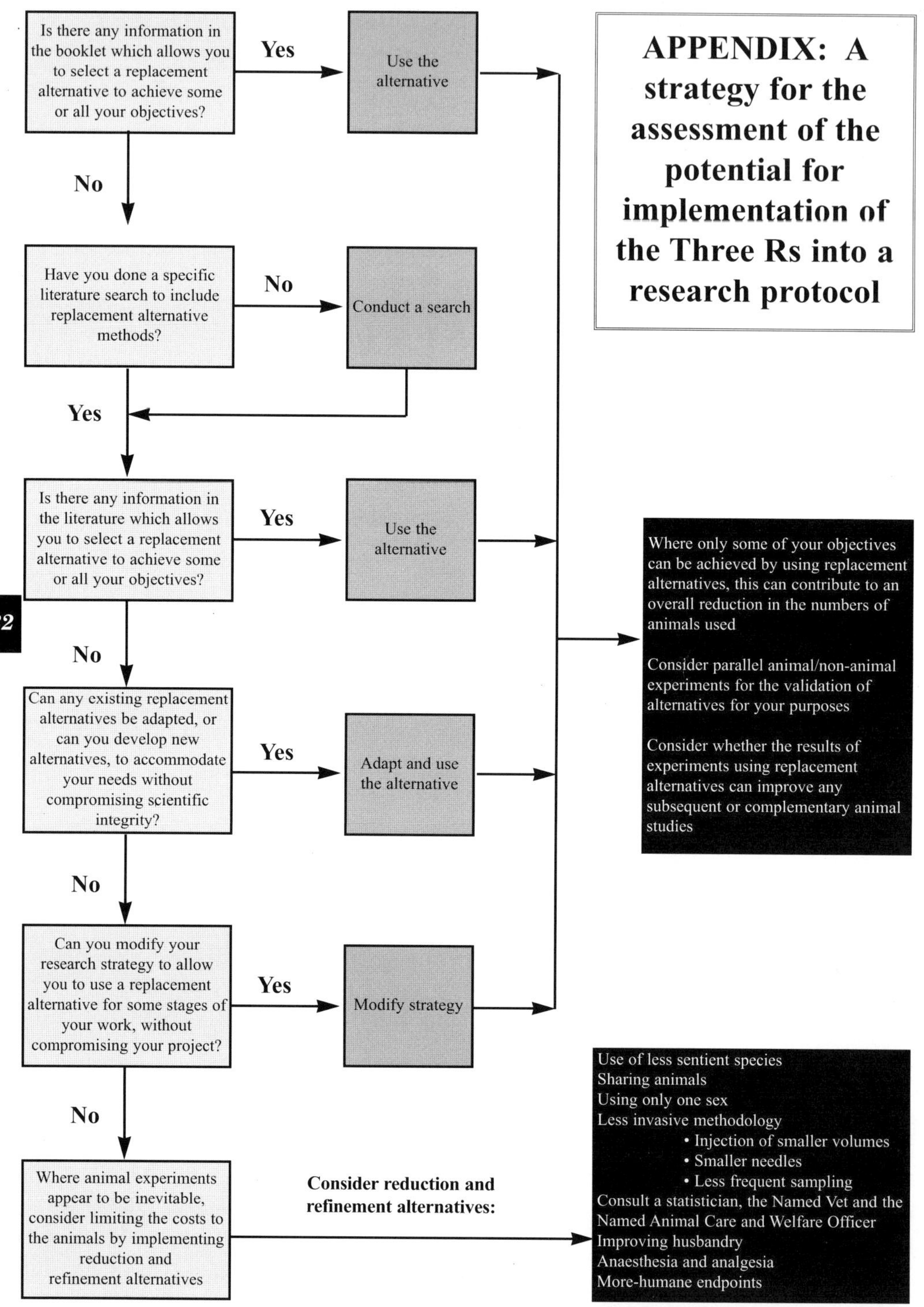